THE STORY OF THE HEAT ENGINE AND THE ORIGIN OF ITS POWER

K.S. WHIPPLE

The Story of the Heat Engine and the Origin of its Power.

LCCN 2022918830

ISBN 979-8-9869749-0-3

CONTENTS

Participants in the Discussion v

1 The Surprising Power of Heat Engines 1

2 Ancient Roots of the Heat Engine........................ 7

3 Something From Nothing.................................... 11

4 Making Heat—the Chain Reaction 23

5 Lavoisier Picks the Lock 35

6 The "Vital Quintessence" 43

7 The Earth and its Role in Combustion 57

Bibliography.. 67

PARTICIPANTS IN THE DISCUSSION

Professor Stanton Bixby

Salviati (Sal)

Simplichio

1

THE SURPRISING POWER
OF HEAT ENGINES

Simplichio — Bixby, I don't understand why you spend so much time yammering on about heat-engines. The principles of operation for these machines have all been known for a long time.

Bixby — For me, Simplichio, it is more intriguing than say, things like quarks or relativity. Modern science is just not that interesting. Heat engines, on the other hand, deal with things that we encounter constantly in our everyday lives. I am speaking of such phenomena as fire, heat, and vacuum, just to name a few. Knowledge of this subject will help us to understand several areas of science.

Simplichio — Perhaps so, Bixby, but heat- engines have a bleak future — they are on their way out.

Bixby — I hope not. I don't think it would be a good idea to jettison a device that has been so important.

Simplichio — Important? In what way? So far, the heat engine has been just a necessary evil.

Bixby — Let's look at it this way, Simplichio. In a modern economy, energy is the straw that stirs the drink. Heat engines have played a vital role in producing that energy. Sadi Carnot, a French scientist, expressed this view succinctly when he said, in 1830, "The study of these engines is of the greatest interest. Their importance is enormous, as they seem destined to produce a great revolution in the civilized world."[1] Yes, Simplichio, not only in the civilized world, but in the entire world. A denizen in our time stands by his vehicle much like the cowboy stood by his horse. Out cities have become habitats as much for cars as for people.

Salvio — I think that the professor is right, Simplichio. The heat-engine is deserving of scrutiny. If the story is not complete, then further rumination on the subject would be valuable.

Bixby — There are also aspects of the engine that I feel are under-appreciated.

---

1   Said Carnot, *Reflections on the Motive Power of Fire*, (New York, Dover Publications, 1968), p3

Simplichio — Such as what?

Bixby — Probably the main one is the density of the energy that the engine produces. It does not produce energy per se, but does so in superlative fashion. Its power is prodigious. I will give you a few examples—with only a gallon of fuel, this engine can propel a family and the vehicle for thirty miles or more. The energy that is produced by this same amount of fuel can run a television for thirty-six days. By weight, this gallon of fuel holds about ninety times as much energy as a lithium battery.

Salvio — Yes, professor, the heat engine is a very impressive machine.

Bixby — Mankind has been changed greatly by it. From the earliest of times, most people have had to earn their living through hard, physical labor, which was usually in the form of agricultural pursuits. The heat-engine has greatly alleviated this situation, and now many people can earn a living in an occupation that does not wear out their bodies.

Simplichio — That is a good point professor, and bravo for the heat-engine. We must, however, keep our eyes on the future.

Bixby — Yes, Simplichio, and in the future, heat engines will play a major role.

Simplichio — I don't see how; their shortcomings are widely recognized.

Bixby — Well, Simplichio, perhaps we should consider those vehicles that will take a starring role for our society in the future.

Simplichio — What, you mean electric vehicles?

Bixby — No, I am speaking of rockets. Those vehicles, transporting us away from planet Earth, are basically cylinders sitting on heat engines. Their propulsion is generated by hot gases being expelled from the rocket after violent chemical reactions have taken place in their engines.

Salviati — I can see that the professor's ideas have substance. Regardless, the more that people understand the activities of nature, the more they feel at home in the world. Technology then, does not seem so dangerous and alien, but can be rather like a curiosity.

Simplichio — You make some good points, professor, and I can sympathize with some of them. This subject, though, is well trod upon, and there is no value in looking at it further. After all, the scientists have been explaining this technology for at least 150 years.

Bixby –Thank you, Simplichio, but I would dispute your contention that the scientists having figured it all out. In 1824, Robert Meikleham said, "Who gave currency to the phrase of the invention of the steam engine being one of the noblest gifts that science ever made to mankind? The fact is that science, or scientific men, never had anything to do in the matter. Indeed, there is no machine or mechanism in which the little that theorists have done is more useless. It arose, and was improved and perfected by working mechanics — and by them only."[2]

Salvio — That viewpoint is surprising. I think that the general perception of technology is that it is a scientific enterprise. This statement, however, seems to contradict that idea.

Bixby — Yes, Sal, and this is another reason why I find the heat-engine and its history so interesting. How much was science involved in its development?

Simplichio — Whatever that is, I'm not sure that it matters. The bottom line is that heat engines are not well liked. They are noisy and pollute the air. Worst of all, in order to procure their fuel supplies, giant drilling machines are deployed. These are used to violate Mother Earth and they cause great harm to the landscape.

---

2   Abraham Woolf, A History of Science, Technology and Philosophy in the 18th Century, (New York, Macmillan and Company, 1952), p611

Bixby — Thank you, Simplichio. I knew I could count on you to provide the salacious details.

Sal — Simplichio is right. The reputation of the heat engine is in tatters. It will take a great deal of work to rehabilitate that reputation and counter the negative opinion that many people have about it.

Bixby — Sal, I agree that it will be a great challenge. I think that most people are content to go to a filling station, dispense some liquid into the tank and embark on their travels without giving the go of the engine a second thought. Perhaps that is where our query should begin. How does this liquid evoke so much power? What is the coin of conversion in this engine?

2

ANCIENT ROOTS OF THE HEAT ENGINE

Salviati — When dealing with the apparatus of the heat-engine, we are confronted with the phenomenon of combustion. It is fire in a jar, so to speak. Can we say exactly, what combustion is?

Simplichio — That's easy to explain, Sal. In this process, we are burning stuff — in particular, fossils. That is why the fuel is called "fossil fuels."

Bixby — Thank you, Simplichio, for that explanation. I was hoping, however, for a more detailed description.

Sal — Let me offer this definition of combustion, professor: When heat is produced in such quantities in a material, then that material begins to glow and eventually bursts into flames.

Bixby — Well done, Sal. The heat engine, then, is a dramatic transformation of this phenomenon—it is a machine that uses fire.

Sal — This ability to create fire goes back a long way. It has been called the most valuable technical skill that Mankind has been able to develop.

Bixby — Yes, Sal, many cultures and mythologies have stories that depict the acquisition of fire for their people—evidence for how important it was to them.

Sal — I vaguely remember the Prometheus story about him stealing the fire.

I guess Western civilization is no exception to the telling of these stories.

Bixby — You would think so, but this is not necessarily the case. I have looked through writings from the Judeo-Christian portion of Western Civilization and have not been able to discover any recognition that the use of fire was of any significance.

Salvio — That is surprising — probably just an unfortunate omission. What about the other main branch, the Greek-Roman?

Bixby — Well, from the Greek perspective, fire was very important — it occupied a prominent place in scientific and philosophical discussions. My favorite account was of the Creation myth of Prometheus that you mentioned earlier.

Sal — Please jog my memory. I don't remember the details of that story.

Bixby — Certainly, Sal. It was told by Plato and recorded in his dialogue, the *Protagoras*. In ancient times, the Earth was under the dominion of the gods of Mount Olympus. A decision was made to populate the Earth with humans and animals. Two brothers, Prometheus and Epimetheus were given this job. Epimetheus took the initiative and started to design and outfit the animals. He gave them many useful qualities, such as size, speed and power. Animals such as deer, elephants and lions, would exhibit these various traits. Other animals were blessed with bodily protections, such as fur coats and thick hides. Others could fly or swim with great rapidity. Epimetheus had equipped the animals very well, and they would be able to survive in a hostile world. Unfortunately for both Epimetheus and the humans, the operation ran into a major glitch. The problem was that there was only a limited number of useful traits that could be doled out to the prospective earthly inhabitants by Epimetheus. He had miscalculated, and when it came time for the humans to be given their attributes for survival, there were none left. With the two brothers in a quandary about what to do to fix the foible of Epimetheus, Prometheus hatched a plan to rescue the humans from their plight of helplessness. The plan was to procure fire and enable humans to utilize it. It was an audacious plan, though, as Zeus, the chief god, had forbidden the use of fire to all but those who resided in the godly realm of Mount Olympus. Prometheus was forced to sneak into the workshop of Hephaestos, the god of the industrial crafts, and pilfer that most valuable of possessions. After receiving this gift, the humans now had their great equalizer and could survive in a dangerous world. Prometheus, as we know, paid a steep price for his transgression. His penalty for stealing the fire was to be chained to a rock and suffer incessant attacks by wild birds.

Sal — Yes, Bixby, this myth does put the phenomenon of fire in a historical perspective. Fire is a critical technic for mankind and, as you say, we now have a mechanized way of using it.

3

SOMETHING FROM NOTHING

Bixby — Most people think of the heat engine as a straightforward mechanical device. As pointed out, the innovations that were made to develop it, were primarily the work of mechanics and other technically minded people. There was, however, one person, who should be singled out in regards to the evolution of the heat-engine. That was the German scientist, Otto Von Guericke, circa 1602-1687. Few people have ever heard of him.

Simplichio — Bixby, you are right, we have not heard of him.

Bixby — Yes, Simplichio, Von Guericke is an obscure name, but he surely deserves to be better known. Historians have generally agreed, that the world of post 1700 was very different from that which existed in the centuries prior and have tried to discern what instigated these major changes. Many have pointed to the accomplishments of two men as being pivotal. The first of these, the astronomer, Nicolas Copernicus, refuted the commonly held belief that the sun rotates around the earth and not the other way around. It was said that Copernicus gave Mankind a new appreciation for scientific enterprises and as a consequence, an impetus for approaching the world more rationally. Unfortunately, the cosmological realities that surround the average person are not that much of a vital interest to him. As far as science is concerned—there is little enthusiasm for this endeavor in a modern society, especially in the field that Copernicus worked in, astronomy.

Salviati — Professor, I would say that your observations ring true.

Bixby — Yes, Sal. The second person whom historians have identified as being of prime importance to the development of the modern world was Christopher Columbus, the Italian seafarer. He is said to have opened up a "New World," and thereby, expanded the imagination of Europeans. While the voyages of Columbus led to huge additions of scientific knowledge and commercial activity, they did not significantly impact mankind in a technological way.

Simplichio — Bixby, you seem to have a cynical attitude towards these giants of history.

Bixby — I did not mean to disparage the accomplishments of these two great men, I only meant to suggest, that the effects of those achievements, were not nearly as momentous as those that have come from the advent of the heat engine. I think that they pale in comparison.

Sal — Professor, you say that the experiments and ideas of von Guericke led to the heat engine being developed. I think we should scrutinize those, but first, what was the initial heat engine and how did it come about?

Simplichio — Bixby, I think I can supply the pertinent information regarding this topic.

Bixby — Yes, Simplichio, go ahead—

Simplichio — The first heat engine was built by Thomas Newcomen in the 18th century. It ran on steam. That is why it was called the "Age of Steam."

Bixby — That is very good research, Simplichio. However, your perception and that of most everybody, that these engines were powered by steam, is off the mark.

Simplichio.-Maybe it is you, Bixby, who is off the mark. It is common knowledge that these engines were steam powered.

Bixby — It is true that steam was produced in the operation of these engines, but the production of power was actually accomplished as a consequence of atmospheric pressure. They were called 'atmospheric' engines at the time, not 'steam' engines.

Sal — Then why the misconception?

Bixby — Modern engines operate on the principle of expanding gases. Inside a jar or cylinder, a chemical reaction takes place that produces these gases. The gases push against a piston creating movement. This movement is a pneumatic spring, similar to a bicycle pump. The atmospheric engine, on the other hand, operated on the principle that two unequal pressures create force. This is where the steam came in. In an atmospheric engine, the steam is condensed or turned into water, by cooling it with water. The space that was formerly occupied with steam becomes a space of water and lower air pressure. This is the result of the change from steam to water.

Sal — I think everybody has probably experienced this phenomenon in the kitchen—when something in a lidded jar has been heated, and then allowed to cool, it becomes difficult to unscrew the lid due to the decreased air pressure.

Bixby — Good point, Sal. With the air pressure on the inside of the cylinder lowered, the air on the outside of the engine, or atmospheric air, was allowed to push down on the piston. The piston assembly was connected by a chain to a rocker arm. The whole apparatus operated much like a children's teeter totter.

Sal — Yes, I can see, the steam was used to produce a dissimilar air pressure and not push directly against a piston. It is surprising that this effect of diminished air pressure, could create enough force to power a workable engine.

Bixby — It is not generally appreciated how powerful atmospheric pressure can be. It has been said that we reside at the bottom of a sea of air—consequently, we have a whole ocean of air on top of us. The pressure from this air presses in on us at force of about 12 to 14 pounds per inch of our bodies. This phenomenon can be shown by sucking the air out of a soda can. The can will crumple by itself just from the atmospheric air bearing down on it.

Sal — In an atmospheric engine, the condensed steam produced a low- pressure area or vacuum. As I understand it, the demonstration of vacuum settled a major controversy in science.

Bixby — Yes, the idea of vacuum had been highly contentious. Since the equipment to make an airless space was unavailable, there was keen debate as to whether it was possible for it to exist or not. Today, the attitude is just the opposite.

Sal — I don't believe that hardly anyone thinks about vacuum today at all.

Bixby — I agree. For modern people, vacuum is a ho hum affair. If one rifles through engineering books, the effect is treated as a very obvious thing—only a bumpkin would not know about it. Physics textbooks take note of it, but in a rather desultory fashion. As I said, it was different in the past. Two of history's greatest minds, Aristotle and Rene Descartes, both asserted, incorrectly, that a state of vacuum could not exist—that space must always be filled with some kind of matter. Therefore, the endeavors by von Guericke and others that showed that a vacuum was possible, was a great breakthrough and led to the development of the first heat- engines.

Sal — And, how was von Guericke able to do this—demonstrate the plausibility of vacuum?

Bixby — It should be pointed out that von Guericke wanted not only to demonstrate this phenomenon, but to exhibit the tremendous potential force that could potentially result. As was explained before, even though the materiality of air seems inconsequential, its cumulative weight is not. Von Guericke reasoned that if a space could be evacuated of its air, a substantial force would be developed on the outside of that space. W. Rosen gave a good description of this process, "The inventive power is inherent in the idea that a column of air has weight in the same way as a column of bricks, and that if some bricks of air are removed from the bottom of the column, the top will move downward."[3]

---

3   William Rosen, *The Most Powerful Idea in the World*, new York, Random House, 2010, pg39

Sal — And, it was with the famous Magdeburg experiment that von Guericke showed this force phenomenon?

Bixby — Yes, Sal. To do this, he first cut a metal sphere or ball into two equal halves. This sphere measured a mere 22 inches across. Von Guericke then fitted them back together using a leather gasket to seal the hemispheres. He had developed an air pump, which was used to remove as much air as possible from the reassembled sphere, resulting in a near vacuum. He then proceeded to exhibit the quantity of force that was pressing in on the sphere from the outside in a rather dramatic fashion. Using hooks and chains, two teams of eight horses were connected to each half of the sphere. Each team was goaded to try and yank the ball apart in opposite directions.[4] They were unable to do so—the pressure on the outside amounted to nearly 3 tons. Consequently, the German scientist had showed that a tremendous force could be generated using only atmospheric pressure. He later constructed a piston mechanism using the same principle and was able to lift a weight of 2600 pounds.[5]

Sal — An astonishing achievement, considering that he was just using potentialities of air. I can see why these experiments would catch the eyes of other scientists.

---

4 Stephen F. Mason, *A History of the Sciences*, (New York, Collier Books, 1962), pg273

5 Thomas E. Conlon, *Thinking About Nothing*, (The Saint Austin Press, 2011) pgs280-281

Simplichio — I fail to see how horses and chains have much to do with how heat engines run.

Bixby — From a practical standpoint, the work of von Guericke was very important. As a result of these experiments, other engineers such as Robert Hooke and Denis Pepin, were inspired to develop technical methods that could harness this power and lead to the development of a workable engine. The principle of vacuum is also critical in the operation of modern compression engines. Mechanically minded people would know that this phenomenon of low air pressure must be utilized in order to induce air into the engine—it would not be operational without it. The ideas of von Guericke were critical for the development of the heat engine, but they also show up in other areas of science as well.

Sal — Such as...

Bixby — Surprisingly, electricity. Up until about 1750, two types of electricity were thought to exist—amber and vitreous. This was because, of course, that it acted in two dissimilar ways. Benjamin Franklin, though, proposed a different explanation for this behavior. He said that it was not two substances that were causing the phenomena, but rather, different concentrations of the same substance. Franklin was discerning in these observations—electrical voltage is now explained as an uneven distribution of charges or electrical pressures; just as an atmospheric engine generated force by dissimilar air pressures.

Sal — Yes, an unexpected correlation.

Bixby — It is in another area, however, where I think that von Guericke makes an invaluable contribution.

Sal — And that is—

Bixby — As I explained before, a primary purpose of von Guericke's work was to develop force to a great degree—in other words he strived to produce concentrated force or energy. In modern times, this phenomenon is taken for granted—heat engines and electrical devices are all around us. This attitude also seems to persist in scientific circles; the subject is only sporadically taken up. Instead of discussions about energy formation or energy densities, the discourse when regarding this subject, usually revolves around such topics such as entropy or energy dissipation. Von Guericke, by contrast, was the first person to emphasize using physical processes to develop concentrated force.

Simplichio — You are wrong, Bixby. You are not making any sense. Everyone knows that there were such things as machines using levers, water wheels, and drill presses in the 17th century. Those were things that produced, as you say, "concentrations of force."

Bixby — Simplicio, that is an excellent observation and valid criticism. Science always needs comments that correct. What I meant to say was that von Guericke was the first person to develop concentrated force or energy using non-corporeal means.

Simplichio — You have me on that, Bixby. I have no idea what you are talking about.

Bixby — That is understandable, Simplichio. What I mean by non-corporeal force is that the force is developed without using something that is bodily or material. Von Guericke did not need the hardware of screws or levers or the physicality of water to produce the effects—the manipulation of air pressure alone, was enough to generate tremendous forces.

Sal — Yes, I would agree with you, professor. That is a different phenomenon altogether.

Bixby — The experiments of von Guericke led to the technic of the heat engine—the way that mankind engages in his everyday activities would now be changed. Utilization of the heat engine led directly to the advent of automation, since a prime mover was now available that could provide a source of power for the machines. Sadi Carnot pointed out another important attribute—on top of the engine's great power, these devices also brought great convenience. They could be in operation seven days a week and not be affected by weather or other environmental conditions.[6] Human and animal labor could now be replaced by automated machines. When von Guericke succeeded at producing a vacuum, he is reputed to have said, "I have seen God." I have not been able to verify that he actually said this, but it would have been apropos.

---

6   Said Carnot, Reflections on the Motive Power of Fire, Mineola, N.Y., Dover Publications, 1988, p3

4

MAKING HEAT—THE CHAIN REACTION

Salviati — We have gotten a general idea that the heat engine is associated with the atmosphere. So far, any discussion of heat is lacking.

Bixby — I assure you Sal, we will make up for that. We needed to explain the work of von Guericke because he discovered properties of gases that are pertinent to our story. Primarily, heat engine operation is based on the medium of gases. This is probably not apparent to the average person whose experience of the fuel requirements for the engine consists of operating a liquid pump at the filling station. An illusion develops, then, that engine operation is dependent on a liquid, and obviously, this is a misconception. In fact, the critical principle of heat engine operation is that of manipulating gases. Professor Konrad Krauskopf, University of California, provided an explanation of this process, "The only practical method ever suggested for obtaining mechanical energy from heat, the method used in both steam and compressed engines, is to supply heat to a compressed gas and let it expand against a piston or the vanes of a turbine."[7]

Sal — That is a very useful and succinct statement.

Bixby — Yes, and I would add that this effect of gas expansion is the only method we have for producing energy density—even nuclear power plants use it.

Sal — There does seem to be a limit on our current technology if we account for these practicalities.

Bixby — I agree Sal, utopia does not seem to be right around the corner. Krauskopf's statement also implies that we must have a grasp of the idea of heat in order to understand heat engine operation.

Sal — This concept of heat—it has been a conundrum of science for a long time.

---

7   Konrad Krauskopf, *Fundamentals of Physical Science*, New York, McGraw Hill, 1959, p146

Bixby — Yes, it has been, and it was not until the beginning of the 19th century, that scientists began to get a decent understanding of what it is. Even today, the subject is a little murky—I think it would be highly beneficial for our cause if we could define it better.

Simplichio — Allow me professor, I will be happy to supply you gentlemen with an explanation of what heat is.

Bixby — By all means, Simplichio, please proceed, we are all ears.

Simplichio — Professor, Salviati, this knowledge has been near and dear to my heart for a very long time. It has come down to us from several geniuses of the past, who through constant and indefatigable labor, were able to gain this knowledge......

Bixby — Yes, Simplichio, please go on.

Simplichio — The whole process of the energy generation starts with the sun, that glorious orb of the solar system. Our sun emits innumerable rays of sunlight to the earth and they are gratefully received by the plants of the earth. This solar energy gets stored in the plants and then, after many millions of years of growth and decay, is transformed and stored in various kinds of hydrocarbons. Consequently, the vestiges of ancient sunlight are available for our use through these hydrocarbons.

Bixby — The hydrocarbons are storing energy?

Simplichio — Yes, yes, Bixby. By George, you are catching on! We need, however, to look at this situation from the atomic level—I am talking about the world of atoms. The energy that is stored, becomes firmly ensconced between the atoms of the hydrocarbons. When the hydrocarbon fuel is run through the engine, the atoms get squeezed.

Bixby — Squeezed?

Simplichio — Yes, it is a bit like squeezing an orange. In the case of the heat engine, we put pressure on the hydrocarbons in the combustion chamber and the result is the production of energy. Unfortunately, the byproducts of this process are not so wonderful.

Bixby — How so, Simplichio?

Simplichio — Well, in the case of making orange juice, all we are left with are the peels and rinds, which of course, are biodegradable. In the case of the hydrocarbons, however, the situation is far different. After the combustion process, we are confronted with remains such as belching smoke, loud noise, and noxious smells. These things were acceptable in the past, but now, we need to go in a different direction.

Bixby — A most magnificent explanation, *Simplichio* — it was very informative. What I most admired in your description is that you offered up a mechanical model of how things work. Most people, when tackling this subject, would throw up a welter of numbers and then call it a day.

Simplichio — Actually, professor, I was tempted to explain this process by just using numbers—and that is my preferred method of explaining things—but, quite frankly, I thought that such a method would have been over your head. You see, I am a firm believer in the cogito of the great Descartes:

Je ai data, donc je existe[8]

Bixby — That is very probably true, Simplichio. I did seem to notice, however, a few gaps in your model. Most especially, was your omission in discussing the formation of heat, or even what it is. Heat is the primary physical result of the engine and for all practical purposes, synonymous with energy.

Sal — Yes, we do need to discuss that part of the subject.

Bixby — People generally conceive of heat as some type of motion. Francis Bacon, early in the 17th century, described it as such.[9] The question naturally arose—what causes the motion? The first major theory suggested that a heat generating substance resided in materials — it was this substance that was responsible for

8   "I have data, therefore I exist"

9   Mason, *History of the Sciences*, p144

heat or combustion. This substance was known as phlogiston(-fluh-gis-ton). The reasoning behind this was that when something burned, it became lighter. The lost material or weight was the phlogiston that had been used. This phlogiston idea later morphed into a similar concept that was referred to as "caloric." John Tyndall, an English scientist of the 19th century, described the theories of phlogiston and caloric as "materialism"—some kind of material was at the basis of heat formation.[10] This principle was challenged decisively by the American, Benjamin Thomson, around the year 1800. He set up experiments to look at how heat was produced by observing the effects of boring out a metal cannon. Thomson astonished onlookers by making water boil by the simple act of the boring(the cannon shaft was immersed in water). No fire or combustible material was needed to produce the heat. He next tested the caloric theory by contrasting a sharp cutting instrument with a dull one. By the caloric theory, the sharper instrument should have produced more heat than the duller one since it would be more incisive in cutting out the caloric. The opposite, however, occurred—the dull blade produced more heat. The modern person would know that the dull blade's production of more heat emanated from the fact that more friction was involved in this operation. Thomson had demonstrated that heat could be produced solely by mechanical means and was not reliant on some reclusive substance or magical elixir.

---

10   John Tyndall, Heat as a Mode of Motion (New York, D. Appleton and Co.,1883), p38

Sal — Since we have ruled out that neither heat nor energy; is dependent on using a substance, what can we say about the heat engine, since we are obviously using substances?

Bixby — Sal, we have to go in an entirely different direction to explain this phenomenon. The instigation of heat in an engine or fire is done by what is referred to as "chemical energy." This chemical energy, however, bears a resemblance to the earlier theories of phlogiston and caloric. Therefore, we should not be too critical of these earlier attempts to account for heat by making a substance responsible for it. Chemical energy is rather quirky and produces surprising results. We should not be surprised, then, if scientists made some rather eccentric speculations regarding it.

Sal — What is quirky about chemical energy?

Bixby — It is this way because during the chemical process, a small cause leads to vast effects. John Tyndall said, "To account for the propagation of fire was one of the difficulties of the last century. A spark was found sufficient to initiate a conflagration. The effects have seemed beyond all proportion to the cause and herein lay the philosophical difficulty."[11]

Sal — I guess I am still in that philosophical difficulty of the 18th century, Bixby. How does a conflagration proceed from a spark?

---

11 Tyndall, Ibid, p66

Bixby — Luckily, Sal, we now have a better idea of what is going on than Tyndall did when he wrote this. The chemical energy is produced by a phenomenon of nature that is called the chain reaction.

Sal — Please elaborate.

Bixby — In a typical combustion reaction, there are two substances that make up the reaction. This would typically be a hydrocarbon and oxygen. They are in the form of a molecule which means that they are in assemblages of more than one atom. These atoms are generally modeled as little balls. The important feature of these molecules is that the atoms are linked together or cemented by "bonds." When these bonds are broken by heat such as fire or a spark, new substances are formed from the separated atoms. When this occurs, heat is given off as a byproduct. The heat is then transferred to the next set of atoms and this operation, called a chain reaction, continues on.[12] It is a bit like knocking down a row of dominos. The first domino initiates the process and then the whole row gets knocked over. One might also liken a chain reaction to a battery of firecrackers.

Sal — And so, this shows once again that heat energy is not generated by a single substance, but is rather as Tyndall says, a 'dynamical' process.

---

12   Mitchell Wilson, *Energy*, (New York, Time Inc. 1963), p93

Bixby — Yes, Sal, in contrast to the lumpen portrayal that is usually given to combustion activities, nature presents us with a remarkable choreography of chemical interactions. An example of this is the hydrogen part of the reaction. It burns first as we can see from the fact that an incomplete combustion leaves us with just the carbon and oxygen components—carbon monoxide.[13]

Sal — The heat produced by the chemical energy imbues the air vapor in the engine?

Bixby — That is right and let us repeat the observation of Professor Krauskopf—only by using expanding gases in a mechanical apparatus, can heat be created in an efficient way. A gas is a quantity of particles that are suspended and moving freely. Imagine a space where these particles are whizzing around like bullets. When the bullets collide or ricochet off the walls of a container, heat is created. This is the heat of friction and consequently, the faster the bullets go, the more heat that will be generated. Also, compression of the gas will result in less space and therefore, more collisions of the particles. This is a coin of conversion, where a gas absorbs energy and pushes against some type of resistance like a piston.

Sal — It is a bit of chaos in the combustion chamber.

---

13   Thomas Gold, *The Deep Hot Biosphere*, (New York, Springer Science Media, 1999), p89

Bixby — Yes, Sal, it is like a demolition derby of particles—the heat is generated by the crashing and banging into each other and the wall of the arena.

Sal — Heat is always said to go from hot to cold, how can we explain this using your model?

Bixby — When two areas are of different temperatures, they try to equal out. What occurs is that the faster particles impart motion to the slower ones, making them gain speed or heat. One could get an idea of this by considering the phenomenon of wind. Because the earth is a sphere or ball, the sun's rays will hit at different intensities along the surface. As a result, there will be a difference in temperature amongst the particles of air in our atmosphere and they will, as a matter of course, try to gain an equilibrium or become similar in temperature. The wind, therefore, is a phenomenon of heat.

Sal — Getting back to chemical energy, I can now see why it has been so useful.

Bixby — Yes, it is quite productive. I would like to point out two other attributes of chemical energy. The first is its functionality in cold weather. Anyone who has experienced frigid temperatures would likely have marveled at how fast and effortless a heat engine can introduce warmth into a space. The technic of breaking bonds renders the engine oblivious to all but the harshest of environmental conditions.

Sal — Yes, the heat engine adds much comfort to people who live in cold places.

Bixby — The second attribute is the matter of convenience. Since chemical energy is a dynamical operation, it can generally only be activated by a willful process. The chemical components are inert until this process is initiated.

Simplicio — Bixby, I don't think it makes any difference, whether or not the chemical parts are hurt.

Bixby — Inert, my good friend. It means that nothing happens until the components are in a reaction. This can be contrasted with a high capacity battery where the energy is always held in high tension.

Sal — Yes, professor, I can see where this convenience would be popular. It is also safer and more efficient to have an on-off switch when dealing with energy.

Bixby — The dynamical process that heat engines use has been very beneficial to Mankind. Unfortunately, confusion has reigned and the production of the heat has almost always been attributed to materialist causes. This was certainly true in prior centuries as I have explained, and continues to this day.

Sal — What do you mean — to this day?

Bixby — Sal, the historical perceptions of what produces heat energy have resembled a game of whack-a-mole. The first great idea of heat creation said that heat was issued from a substance called phlogiston. After that, the critical material responsible for heat energy was identified as caloric. One materialist theory or mole is shot down, but another springs up. Today's mole is hydrocarbons.

5

LAVOISIER PICKS
THE LOCK

Bixby — As the seventeenth century drew to a close, Isaac Newton and his theories had produced a system of physical science that consisted of only two things — matter and motion. Nature and biological life had been reduced to a set of billiard balls that bounced around in mathematical patterns. This was to be a resilient model, as it was succeeded in the twenty-first century by data 'analysis' and these biological entities were now flotsam of some computed algorithm. Chemistry, meanwhile, was in disarray. This was despite it being the most ancient of sciences and the one that people practice the most.

Simplicio — 'The people,' practice chemistry the most? I am not a chemist, nor do I know many people who are.

Bixby — I am speaking figuratively, Simplichio, not literally. If you would consider such activities as cooking, baking, pottery making, metal working etc.; their essential nature is to effect some type of chemical transformation. Therefore, the average person participates in activities of chemistry and has been for a long time.[14] The formal science of the professionals is built on the knowledge and methods that were developed by average people of centuries past.

Salviati — The formal structure and principles of chemistry didn't come into existence until late in the 18th century. This was quite a bit later than when Galileo and Newton were establishing the techniques and subject matter of physics.

Bixby — The wait was worth it, however, as it was the knowledge of chemistry that enabled the industrial revolution to take place. Physics, on the other hand, had mainly been used to calculate how to lob artillery shells accurately at people who had not been born in your locale.

Simplichio — That is ridiculous, Bixby. Physics has been used to explore the heavens and was just then starting to understand electricity.

Bixby — That is just hyperbole, Simplichio, but I welcome your critical attitude.

---

14 Douglas McKie, *Antoine Lavoisier*, (New York, De Cao Press, 1952), p34

Sal — What were the difficulties in setting up chemistry as a formal science?

Bixby — The main problems were twofold. The first problem was that the main components of chemistry — the elements — are hardly ever found in their natural state in the world. They are not laying on the ground somewhere, waiting for someone to pick them up and examine them. Rather, they are almost always hidden by virtue of being part of a mixture of something else. Iron, for example, is usually found as part of a rock like substance and has to be pains-takingly separated from the rock in order to be useful. With regard to oxygen, a main component of combustion, *The Elementary Study of Chemistry*, has this to say: "while oxygen is very abundant, it does not occur in nature in pure condition. To obtain pure oxygen, we must either liberate it from some compound or separate it from the gases with which it is mixed in the air."[15] The first great challenge for chemistry, therefore, was just to identify the materials or elements that you were working with. The second difficult problem of chemistry was the lack of recognizable or sequential patterns among the phenomena. An example of this was the previously mentioned observation on fire — a tiny spark could ignite a large fire or conflagration. A small thing would cause a large effect. The seeming illogical properties of fire concerned how and why it is extinguished. This, of course, can be done by dousing the

---

15   William McPherson, William Edwards Henderson, *An Elementary Study of Chemistry*,(Boston, Ginn and Company,1933), p33

fire with water. The explanation for why this happens becomes problematical, however, when one considers that the water is made up of two inflammable gases — hydrogen and oxygen. The plausible guess would be that these two elements in tandem would increase the fire, not put it out. A possible resolution to these paradoxes was put forth by Robert Boyle in the seventeenth century. He suggested that the elements or main substances were like letters of an alphabet. Just as using letters in different combinations could produce different words and meanings, so might changing the orders or amounts of substances in a chemical composition, produce new substances and effects.[16] Even a small adjustment could evoke an entirely different phenomenon.

Simplichio — We seem to be getting far afield from a discussion on heat engines.

Bixby — Sorry about that Simplichio, but when one considers that oxygen plays a major role in combustion, I thought it would be valuable to provide a background to its identification as an element.

Simplichio – Go ahead. By all means.

Bixby — The momentous discovery of oxygen is generally presented as a cut and dried affair that involved Joseph Priestley and Antoine Lavoisier. The reality was far different.

---

16   Mason, *History of the Sciences*, pg 239

Sal — How so?

Bixby — An historian, J. W. Severinghaus,[17] reports that as many as eight people in the space of five centuries had speculated that 'there was something going on in the air,' and that something, had a connection to the phenomenon of fire. It was, however, in the seventeenth century, that scientists made some remarkable observations regarding oxygen and combustion. In 1660, Robert Boyle found that when he removed air from containers in which he had placed fire or mice, the fire went out and the mice died. Robert Hooke, in his famous book of 1665, the *Micrographia*, asserted that a substance in the air was active in the process of combustion.[18] He was unable to verify this experimentally. John Mayow, (1640-1679) of England was more successful. He conducted experiments that showed a diminishing of air from a mouse's respiration and from a burning candle. In 1674, he postulated that the increase of weight that came from the burning of metals was caused by a reaction with 'nitro-aerial' particles in the air.[19]

Sal — This is not widely known history.

-----

17   J.W. Severinghaus, *Revisionisms of Oxygen's Discovery*, Adv Exp Med Biol., 2003-543

18   McKie, *Lavoisier*, p42

19   Ibid, pp.gs 44-45

Bixby — Yes, Sal, cogitations on the nature of air had led scientists of the seventeenth century right to the brink of discovering what oxygen is and the recognition that it has a role in combustion. And yet, this knowledge was basically lost as science went into a general hibernation regarding these speculations.[20] This lull in conjecturing

about oxygen lasted for about a century and demonstrates that science is not always a steady march forward.

Sal — The challenge of identifying oxygen does seem to have been an arduous task. Then, it was in the 1770's, that the breakthrough actually occurred?

Bixby — Yes. In 1778, Joseph Priestley isolated some pure oxygen over a basin of water and was entranced by its properties that could somehow augment a flame and enhance respiration. Carl Scheele, an apothecary in Sweden, conducted similar experiments but failed in his attempt to communicate his results to the world in a timely fashion. Although, Priestley had conducted the experiments to isolate oxygen, he was unable to articulate an explanation that would account for the element being a causative agent of combustion. He still supported the theory of phlogiston. Priestley had relayed the results of his work to Antonin Lavoisier and they had intrigued the French scientist greatly. Lavoisier had been perfecting the practice of meticulously weighing the air that had been

---

20   Ibid, pg 49

produced in various reactions. This was rightfully seen at the time as a rather bizarre methodology. He gained success, however, when after observing that a metal became heavier after being heated in a flame, he concluded that the metal was reacting with a constituent of the air. The reacting gas was a primary element and was given the name, oxygen. The French scientist had lifted the blinders off Mankind's eyes and a new era of knowledge about combustion began.

6

THE "VITAL QUINTESSENCE"

Simplichio — We have spent all this time going on and on about heat engines only to find out that oxygen was discovered — are you going to explain to me next that the sea is salty?

Bixby — I can sympathize with your impatience, Simplichio, but this telling of history was necessary since chemistry and oxygen play a major part in our discussion.

Salviati — Yes, many people say that the role of oxygen in heat-engine operation is greatly undervalued.

Bixby — I would agree, Sal. Let us discuss this further. Many people get their idea of how fire and oxygen work from the common camp-fire—this reaction involves the carbon of wood being mixed with the oxygen that is contained in the ambient air. In a heat engine, however, the oxygen functions very differently. The piston assembly acts as a high -powered pump, drawing in huge quantities of oxygen to feed the engine. To better understand this activity, we should as a matter of course, determine the quantity of substances that go into its operation.

Simplichio — Bixby, I can handle this. Numbers are my forte. Of course, the heat engine consumes large quantities of air—at a speed of 45 miles per hour, the average auto takes in about 1800 liters of air per minute. This translates into a ratio of 14 parts air to 1 part hydrocarbons. Obviously, it takes a lot of air to aerate and support the combustion.

Bixby — Thank you, Simplichio, but I do see a minor error in your calculations.

Simplichio — And that is?

Bixby — You give figures that reference 'air' — but in actuality, only about a quarter of the air consumed by the engine is used in the combustion process. I know that engineering books like to use these numbers but they are confused about which substances react and which do not.

Simplichio — And how did you arrive at this conclusion, good Professor?

Bixby — To understand the process, we must take into consideration the composition of the atmosphere—the air that is downloaded into the engine consists of 78% nitrogen, 21% oxygen, and 1% miscellaneous gases. Only the oxygen reacts with the hydrocarbons, so any calculations that utilize 'air' — would be erroneous.

Sal — I guess the logical question would be to ask, what the proportions are of each of the participating entities — the hydrocarbons and the oxygen?

Bixby — Yes, Sal, that is very valuable information. To accomplish that, we need to employ a chemical equation.

Sal — But Professor, isn't it necessary to be a trained chemist or at least a student of chemistry, to understand a chemical equation?

Bixby — No, that is not necessary. A chemical equation uses simple arithmetical ratios—a methodology that most can accomplish by the third grade. The ratios are based on the weights of the individual elements. In this case, oxygen$(O) = 16$, carbon$(C) = 12$, and hydrogen$(H) = 1$. Let me give an example—this is for an octane reaction, which is a type of gasoline:

$$25O_2 + 2C_8H_{18} \rightarrow 16CO_2 + 18H_2O + heat$$

We disregard the right side of the equation since we are solely interested in the reactants of the left side. For oxygen, the quantity is 25x(16x2)=800 units. For hydrocarbons, the quantity is 2x(8x12)+2x(18x1)=228 units. These numbers tell us that for an average combustion reaction, the compositional weight of the reactants is 22% hydrocarbons and 78% oxygen.

Sal — I can see that the presence of oxygen is quite prominent. I would ask, what are the properties of oxygen? What is it like?

Bixby — That is a good question, Sal. Our everyday experience of breathing in oxygen is a little deceptive because the oxygen that we inhale is buffered by its low concentrations in the air overall. As we explained before, air is made up of only 21% oxygen. Seemingly innocuous and pleasant, the reality is that oxygen is a very volatile and caustic substance.

Simplichio — Caustic? I didn't bring my dictionary. What is it?

Bixby — This is the word that Lavoisier, himself, used to describe the nature of oxygen. It means that it is corrosive—a substance that will burn or eat away another substance. Robert Hooke, the 17th- century English scientist, had a similar view of oxygen when he said, "the air in which we live, move and breathe....this air is the menstruum or universal dissolvent..."[21] This quality of oxygen, which corrodes or dissolves substances, is dramatically demonstrated by the phenomenon of rust—a common everyday

---

21   Robert Hooke, *Micrographia*, Chapter 16

occurrence which is the result of a material being enveloped in a gas mixture containing oxygen—and having the oxygen burn that material. This process has been classified by the science of chemistry as an oxidation reaction and it has been said that the damage to property from rust surpasses fire.

Sal — What other things demonstrate this aspect of oxygen?

Bixby — The Elementary Study informs us that oxygen is used in sewage wastewater operations. Putrid water is sprayed into the air where the oxygen promotes the cleansing of the water simply by making contact.[22] Water is another example that shows how active oxygen can be, chemically speaking.

Sal — How is that?

Bixby — Water is generally thought of as a soft and benign substance. When viewing impressive rock formations that have been created by water erosion, some geologists will tell you that these occur despite water's supposed innocuous properties. Because water contains dissolved oxygen, it behaves as a corrosive—so it should not be surprising to see dramatic effects caused by water movement. With this corrosive quality of oxygen, one could conclude that in order to clean something to a certain extent, it

22 *Elementary Study*, Ibid, p40

is only necessary to expose that something to air or water and therefore, oxygen.

Sal — Do you have any other examples, Professor?

Bixby — Oxygen is a volatile substance. If one were to drop a burning ember into a test tube of oxygen, a brilliant flame would result. Then, there is the illustration of granaries(buildings that store grain) — vigilance must be maintained to control amounts of oxygen-too much exposure of grains to oxygen could result in a spontaneous fire.[23] This same scenario plays out in the combustion with hydrocarbons. Natural gas is the most efficient mode of combustion because it is easier to envelop the gas particles with oxygen. Liquid hydrocarbons are the next most effective way as the liquid needs to be made into a spray before the combustion process is initiated—the least efficient is coal, as these rocks are not homogeneous with a gas medium such as oxygen.

Sal — These qualities of oxygen are not quite what I expected.

Bixby — Yes, Sal, but many of these things are phenomena that can be observed visually. The situation with an engine, however, is far different—we are not able to see the operation directly and so we may be misled as to what is actually going on in regards to the

---

23   Ibid

participating substances. This points to another important feature of oxygen that is not generally appreciated.

Sal — What is that?

Bixby — The primary process of the engine operation is probably that the oxygen hydrocarbon mixture is compressed to a great degree—this raises the temperature of the mixture to about 1500 degrees Fahrenheit.

Sal — And what is the consequence of this activity?

Bixby — Well, with respect to the oxygen, you have a completely different substance when this compression occurs. Let's let the Elementary Study explain, "At ordinary temperatures, oxygen is not very active, most substances either are not affected by it or are affected slowly. At higher temperatures, however, oxygen is very active and unites directly with most of the elements."[24] This compression is greatly abetted by an undervalued component of the engine—the flywheel. It stores mechanical energy from the engine and uses this energy for compression of the fuel mixture. The continuous compression of the mixture serves to multiply the energy production of the engine.

---

24   Ibid, p34

Sal — Yes, professor, I am beginning to see why oxygen is categorized as a flammable gas.

Simplichio — Well, I fail to see to see the value of these comments. It is starting to sound like you guys think that oxygen plays a major role in the production of heat energy. We know that this is not true and the science has been settled on the topic for quite some time—oxygen merely supports the combustion process.

Bixby — Simplichio, I am aware that your description is the prevailing one. Perhaps, though, I might change your mind if I brought to your attention, the case of the 19th century cigar lighter.

Simplichio — The 19th century cigar lighter? Bixby, it sounds like you are going off the rails.

Bixby — Perhaps so, good fellow. What I am referring to is an episode in engineering that took place in Germany in the 1890's. I hope it will help to illustrate the point that I have been trying to make.

Sal — Please proceed, Professor, I am anxious to hear about this.

Bixby — At that time, a German engineer named Rudolf Diesel, had been immersed in the design of the most powerful of heat-engines — the diesel. In the midst of his labors, he had come across this cigar lighter—a device that operated in a rather curious way.

Sal — And that was?

Bixby — The lighter produced a flame to light the cigar by the simple process of compressing air in a cylinder by using a miniature piston. It was referred to as a pneumatic lighter.

Sal — Yes, I think I may have seen one of these somewhere.

Bixby — They are still available today, *Sal* — a person can sometimes purchase one at a sporting goods store. They are also called 'fire syringes' since they function in much the same way as their medical counterpart. The syringe is simply pressed down to produce a flame. Being inspired, in part, by the fire syringe, Diesel used the principle of gas compression as the main operational feature of his engine. For example, Diesel's engine does not use spark igniters to start the engine, but relies solely on compressing the hydrocarbon oxygen mixture to start the engine.

Simplichio — So what? There is still a hydrocarbon component in the mixture of diesel fuels.

Bixby — Yes, Simplichio, but if we go back and consider the operation of the cigar lighter, we see that combustion is occurring without that hydrocarbon component. We must conclude, therefore, that the oxygen contained in the air, by itself, can produce combustion. The idea that oxygen is just an adjunct to this combustion, does not square with what we observe in the cigar lighter operation

— the assistance of a carbon substance is not necessary to generate a flame.

Simplichio — *Sacre bleu*, Bixby!—This cannot be!—Oxygen is just a support in the process!

Sal — But Simplichio, it does seem that the professor is presenting evidence that challenges your assertion that oxygen is just a support. Professor, does this issue of defining the role of oxygen in combustion as either a primary component or just an adjunct, have an historical background?

Bixby — Yes, Sal, as a matter of fact it does. It was addressed by John Tyndall.

Sal — Tyndall was a Victorian scientist, whom we have cited before. He helped to pioneer the concept of the atmosphere absorbing heat from terrestrial sources.

Simplichio — The "father of global warming?"

Sal — Yes, the same person.

Simplichio — Okay, I am listening.

Bixby — That is great to hear, Simplichio, and I hope that you will stay interested after I present this information on the subject. Mr.

Tyndall describes the process of combustion and uses the example of coal gas since this was the predominant hydrocarbon in use at the time(middle of the 19th century). He says this: "We sometimes hear coal gas spoken of as a combustible substance, and oxygen as a supporter of combustion, but if our atmosphere were composed of coal gas and if the great gas works were filled with oxygen, we should be able to burn that oxygen in an atmosphere of coal gas. Or if, instead of being filled with oxygen, the gas holders were filled with common air, we should be able to burn the air.[25] So, you see, gentlemen, the two main components of combustion are interchangeable. That is to say, as Tyndall categorized it, the combustion process is dynamical and not materialist.

Sal — Well, professor, I think you have demonstrated that the production of heat energy cannot be ascribed solely to the presence of hydrocarbons—the dogma of a materialism based on these hydrocarbons is very limited in its ability to explain combustion. You have previously provided numbers that show that oxygen is the main feedstock in the process—about three fourths of the substances used. While hydrocarbons are, of course vital in this activity, prominence should also be given to the oxygen component. Why, professor, would you say that the materialist explanation of combustion has persisted for so long?

---

25   John Tyndall, *Heat as a Mode of Motion*, p66

Bixby — Sal, the first reason for this attitude, I would say, is the visual. As I have noted before, we observe liquid being dispensed at the filling station and we know from experience that the engine will not run without it. What is not taken into consideration is the hundreds of pounds of oxygen that is gulped down by the engine in the process of consuming that liquid. It is an optical illusion—a bit like when we observe the motions of the sun and earth. During the day, the sun starts in the east and slowly passes over to the west. It appears as though the sun is going around the earth. The second reason is human nature. The hydrocarbons come to us only after painful extraction and cost. The predominant substance in the combustion process is oxygen, but since this a free component, little consideration is given to it. The third reason is that the heat engine was mostly developed by technologists(engineers) and mechanics. It was not imperative for anybody to go into the whys and wherefores of how the energy was produced. As Thomas Gold said, if something works, there is not a binding necessity to explain it. The fourth reason is language and its use.

Sal — Language?

Bixby — Strangely enough, yes. The technical language that has been built up for chemistry is adept at handling large amounts of information in an economical way. It is lacking, however, in terminology that allows average people to understand ideas and concepts in an intuitive way. When Lavoisier set up the linguistic framework for chemistry, the more descriptive language of everyday life was jettisoned. Lost were the more physical, visceral accounts of substances that were made by the avid, hobbyist scientists who discovered them.

Sal — Can you give some examples of this?

Bixby — Yes. When hydrogen was identified by Cavendish in the 18th century, he referred to it as "inflammable air." Carbon dioxide, at that time, was known by the quaint term, "fixed air"—given because it was collected from solid substances. When Carl Scheele first isolated oxygen in 1775, he labeled it as "fire air."[26] This was a concise and clear description of what the substance was actually like and what function it performed. Perhaps the best moniker for oxygen was given by Robert Boyle in the aborted effort to identify the element in the 17th century. Boyle had found that animals expired and flames could not stay lit when placed in enclosures bereft of air. Boyle concluded that the air obviously contained a critical substance that he referred to as the "vital quintessence."[27]

Sal — Yes, professor, that is a fitting description for oxygen. And, as you say, Scheele's name, 'fire air,' might make one stop and ponder its nature.

---

26 Mason, *History of the Sciences*, p340

27 Ibid, p240

Bixby — Perhaps so, Sal—there has been a great deal of misunderstanding on the subject. An example of this was shown by a disastrous fire that occurred during the Apollo moon mission program. On June 1, 1967, three of America's best astronauts died when a fire started in the Apollo One test capsule. The interior space of the capsule had been filled with 100% oxygen when a tiny spark ignited the gas. The lamentable result was that the poor astronauts were incinerated inside the capsule. According to James McDivitt, a veteran astronaut of the program, "We had no idea what the effect of a high oxygen content environment does as far as burning goes and we learned the hard way."[28] The NASA engineers who designed the air compartment of the space capsule were indifferent to the volatility of oxygen but were then forced into a reappraisal. For the rest of Mankind, even today, this quality of oxygen is little recognized—oxygen is seen only as a 'support' in the process of combustion.

---

28 British Broadcasting Corporation, *Nasa: Triumph and the Tragedy, Episode 1*

7

THE EARTH AND ITS ROLE IN COMBUSTION

Sal — Professor, I appreciate your efforts to try to understand why the heat engine is so effective. As we have discussed, the weight of the atmosphere functions as a gigantic bellows—a mechanism that acts to press enormous amounts of air into the engine. This process is based on principles that were developed by Otto von Guericke. Now we see that this air contains the primary fuel by weight for the engine in the form of oxygen. Therefore, even though the term, "atmospheric engine," was affixed to this device at the time of its conception, it is probably a more apt name for it today.

Simplichio — It is terrible, Bixby. You are saying that there is an energy source sitting here, right on top of our heads? This idea is absurd—if that were true, we would all burn up since the oxygen is all around us. It is that simple.

Bixby — That is a reasonable objection, Simplichio, and I will attempt to answer it. Oxygen is a volatile substance that exists in our atmosphere. The reason that we don't burn up, as you say, is based on a mitigating circumstance in that atmosphere.

Sal — And what is this circumstance that alleviates the situation?

Bixby — We need to look at the composition of our atmosphere. You will recall that this composition is about 78%nitrogen and 21% oxygen. It is this makeup of the atmosphere that enables us to utilize oxygen in its many facets.

Sal — And how is that?

Bixby — Sal, I hope you will remember when I explained that the elements of nature almost always exist as part of a mixture or a compound. The elements must be laboriously separated from these compounds, and then put in some type of rearrangement, in order to be used. There is the example of hydrogen, a more effective fuel than oxygen, but an element that has to undergo a costly and arduous separation process before it can be utilized.

Sal — Then, what is the case with oxygen? A separation procedure isn't necessary, isn't this right?

Bixby — As we said, the atmosphere is a mixture of oxygen and the inert gas, nitrogen. It is this inertness that performs the critical function. The nitrogen acts as an insulator to the caustic oxygen and limits its effects. If one were to consider an electrical wire carrying electricity, for example, it is the rubber sheathing around the wire that contains the energy and enables the wire to be a practical transport system. The nitrogen in our atmosphere works in much the same way, limiting the oxygen enough, that conditions are favorable for both biological and technical activities.

Sal — It is surprising that nitrogen plays such a vital role.

Bixby — Thomas Gold has provided a good explanation as to why muted energy sources are so important. He is referring to food, but the same principle applies in the production of heat—" it is important to remember that a chemical fuel is useless to life if it combusts spontaneously. Dinner will do you no good if the food bursts into flames on your plate."[29] We have seen that oxygen is combusting all around us in the phenomenon of rust, but this occurs only in a limited way—the active chemical energy is constrained by the presence of nitrogen which effects a dilution of oxygen's intensity.

Sal — Definitely, this is a fortuitous circumstance. I do, however, have another question.

Bixby — Go ahead.

---

29   Gold, *Deep Hot Biosphere*, p18

Sal — Oxygen is utilized in quantities that are prodigious. How does oxygen get produced and how does this supply get replenished?

Bixby — These are useful questions, Sal. Joseph Priestley asked a similar question in the 18th century, "How does nature repair the injuries that are caused by the instances of combustion?" He was referring to the incessant burning of stuff by both people and nature since time immemorial. When something is combusted, the left-over products are water and carbon dioxide. The water, of course, is innocuous, but what was the effect of the produced carbon dioxide? Priestley had isolated carbon dioxide in an enclosure and then placed plants in that enclosure. Surprisingly, he found that the plants thrived—the carbon dioxide was plant nourishment. Priestley concluded then, that the injuries caused to the atmosphere by combustion, were remedied by the processes of plant metabolism. A similar situation occurs in the formation of oxygen. The Elementary Study describes it thusly, "Under the influence of sunlight, the carbon dioxide that has been absorbed by the plants reacts with water and small amounts of other substances from the soil to form complex compounds of carbon which constitute the essential part of the plant tissue. These reactions are attended by the evolution of oxygen which is restored to the air."[30]

---

30 *Elementary Study*, p181

Sal — And so, the production of oxygen is a result of plant metabolism—this being instigated by the absorption of sunlight.

Bixby — Yes, that sums it up nicely. As we have concluded, the critical substance of combustion by weight, oxygen, evolves by plant life and the solar energy that it has imbibed. In a general sense, one could say that heat-engine operation is the downstream result of solar energy.

Simplichio — These heat- engine vehicles are solar powered? It can't be. I don't see any panels on them.

Bixby — From what I understand, Simplichio, the engineers are working on installing some panels on the vehicles.

Sal — The idea that plants play an important role in combustion processes is surprising.

Bixby — It shouldn't be, but that is the reality. All science minded people know that plants produce oxygen. It is a small step to the recognition that this mechanism of plants is critical to heat engine operation.

Sal — If your ideas are correct, it would seem that nature operates in a very convenient and efficient way—merely from vegetation and the oxygen that is produced, Mankind benefits greatly.

Bixby — It might be expected, though. In my opinion, nature always operates in a minimalist mode.

Sal — The heat- engine would seem to fit that portrayal. Instead of a lot of bric-a-brac such as solar panels and wind turbines, we are able to achieve great energy densities using non-corporeal means. This was initiated, as I have said, by Otto von Guericke.

Sal — What is this principle of minimalism?

Bixby — The main idea is to use the least amount of material possible while achieving the greatest effect.

Sal — Yes, I think that that is how nature generally works.

Bixby — My favorite example of minimalism is how nature has designed bones. You would think that they would have been made solid in order to provide maximum strength, but they are in fact light weight and in some instances, just hollow cylinders. Somehow, nature figures out that the strength requirement for bones does not necessitate that they be thick and heavy.

Sal — Yes, sort of like wind turbines. Professor, some would argue that electric cars exemplify this principle of minimalism. They are contrived with fewer parts than conventional cars.

Bixby — Sal, that is a good point—electric cars do have fewer parts.

Simplichio—Electric cars are the future; I think that is pretty obvious.

Bixby — Simplichio, before we get too enamored with electric cars, we might want to consider the nature of electricity. Electricity is generally thought of as a type of energy — but this opinion is not unanimous. In 1907, for example, Oliver Lodge, an important theorist on electromagnetism, said this, "Electricity may possibly be a form of matter—it is not a form of energy. It is quite true that electricity under pressure or in motion represents energy; but the same thing is true of water and air, and we not therefore deny them to be forms of matter."[31] In other words, electricity is a medium through which energy is passed.

Sal — This is an unconventional view of electricity—physics textbooks always classify electricity as a type of energy.

Bixby — Sal, if you look at the phenomenon of electricity closely, you will notice that when it takes place, it must always be pushed from something else. That something else could be a coal fired power plant, chemical reactions in the sun, or scuffing your foot across a carpet. To summarize, electricity is not spontaneous or a resource that can be tapped.

---

31 Oliver Lodge, *Modern Views of Electricity*, (London, Macmillan and Company, 1907), Chapter 1

Simplichio — I don't think we should get bogged down delving into obscurities of science or sit around playing word games. The fact remains that heat- engines are a menace to society. They are noisy, inconvenient, and belch too much smoke. They should be banned.

Bixby — I am sorry that you feel this way, Simplichio. Especially so, since I may have some bad news for you, my friend.

Simplichio — I guess the bad news is that you are running out of errors for me to correct. That would be terrible.

Bixby — Perhaps, Simplichio, I do have one more. This possible miscue is from Professor Harvey White. It is his description of human beings: "The human body may be considered as a heat-engine in which food and oxygen are taken in as fuel, external work is done by muscular activity, and heat and carbon dioxide are exhausted as the principle waste products."[32] Simplichio, I am sorry, but I must conclude—you are a heat- engine! And, we sure wouldn't want to ban you, my friend, would we?

Simplichio — Grrr... Bixby...you could say that I am a heat engine but that is a mere technicality. The bottom line is that I want a clean and peaceful world—the use of your heat-engines does not fit into those criteria.

---

32 Harvey E. White, *Modern College Physics*, (New York, D. Van Nostrand Company, 1955), p319

Bixby — I appreciate those sentiments, Simplichio. However, let me make one final point.

Simplichio — Go ahead.

Bixby — It was Antoine Lavoisier; himself, who made the following observation: the activities that give Mankind its energy are not ordinary activities. That is to say; that in order to produce vast quantities of energy, it takes something extraordinary.[33] Simplichio, if the resources that you claim are so easily accessible and all around us, were sufficient to make things and move them, the presence of those resources, would make the earth uninhabitable. In other words, sunlight and wind are necessarily weak and these conditions provide Mankind and the rest of biology with a bucolic environment from which they can thrive. If sunlight and wind were very powerful, they would turn the earth into a hellscape and not very conducive to life, or us, having this congenial conversation at the present time. And so, we are lucky to have this device called the heat-engine and thereby, be able to procure necessary large amounts of power. Its operation can be confined to a localized place and that is very beneficial. Simplichio, I wish that you would consider these points that I have made.

Simplichio — I will think about it.

---

33 Antoine Lavoisier, *Elements of Chemistry*, (New York, Dover Publications, 1965), p415

Sal — Professor, trying to understand what is behind the heat engine is not easy. This undertaking seems to be like grappling with some sort of scientific Gordion Knot –its operating principles take equal parts from technology, physics, and chemistry. I think you have succeeded somewhat in making this subject less perplexing. Thanks for sharing your ideas.

Bixby — And thanks to you gentlemen, for listening.

BIBLIOGRAPHY

Sadi Carnot, *Reflections on the Motive Power of Fire*, Dover Publications, Mineola, N.Y., 1988

Thomas E. Conlon, *Thinking about Nothing*, The Saint Austin Press, 2011

Thomas Gold, *The Deep Hot Biosphere the Myth of Fossil Fuels*, Springer, Science+Business Media, New York, 1999

Konrad Bates Krauskopf, *Fundamentals of Physical Science*, McGraw Hill, New Yorf,1959

Antoine Lavoisier, *Elements of Chemistry*, Dover Publications, New York, 1965

Stephen F. Mason, *A History of the Sciences*, Collier Books, New York, 1956

Douglas McKie *Antoine Lavoisier*, Da Capo Press, New York, 1952

William McPherson, William Edwards Henderson, *An Elementary Study of Chemistry*, Ginn and Company, Boston Ma.,1905

John Tyndall, *Heat as a Mode of Motion*, D. Applelton and Company, 1883

Harvey E. White, *Modern College Physics*, D. Van Nostrand Company, Princeton, N.J.,1948

Mitchell Wilson, *Energy*, Time Incorporated, New York, 1963

Abraham Wolf, *A History of Science, Technology, and Philosophy in the 18th Century*, Harper and Brothers, 1961